MW00562006

FROM **CEMENT** TO **BRIDGE**

by Robin Nelson

Lerner Publications Company / Minneapolis

Special thanks to: Mark Spafford, Daniel Dorgan, and staff at the Minnesota Department of Transportation and Ronald Medlock at the Texas Department of Transportation.

Lerner Publications Company
A division of Lerner Publishing Group
241 First Avenue North
Minneapolis, MN 55401 U.S.A.

Website address: www.lernerbooks.com

Library of Congress Cataloging-in-Publication Data

Nelson, Robin, 1971–
    From cement to bridge / by Robin Nelson.
        p.   cm. — (Start to finish)
    Includes index.
    Summary: Describes how cement, metal, concrete, and other materials are made into a bridge.
    ISBN: 0–8225–1389–7 (lib. bdg. : alk. paper)
    1. Bridges, Concrete—Design and construction—Juvenile literature. 2. Cement—Juvenile literature. [1. Bridges—Design and construction. 2. Building materials.] I. Title. II. Start to finish (Minneapolis, Minn.)
TG335.N35 2004
624.2—dc22   2003011734

Manufactured in the United States of America
1 2 3 4 5 6 – DP – 09 08 07 06 05 04

Additional photographs in this book appear courtesy of: © Photodisc Royalty Free by Getty Images, front cover; © Brand X Pictures by Getty Images, p.23.

Table of Contents

The cement is mixed 4

Workers prepare the land. . . 6

Workers make forms 8

Workers build the ends. . . 10

Workers build the legs . . . 12

A machine lays beams . . . 14

Workers place bars 16

Workers make the bridge's deck 18

Workers add fences and lights 20

Cars drive across the bridge 22

Glossary. 24

Index 24

Bridges let us cross.

How is a bridge built?

The cement is mixed.

Cement is made from stone and clay. Workers pour cement, water, sand, and small stones into containers on trucks. The containers spin slowly. The mixture turns into **concrete.** Concrete is used to build buildings and bridges. The trucks take the concrete to where the bridge will be built.

Workers prepare the land.

Workers put up concrete fences to protect passing cars. Then workers dig holes and fill them with rock. Workers will make concrete blocks on top of the rock. They will build the bridge on these concrete blocks.

7

Workers make forms.

Forms are hollow containers. They help shape the concrete that holds up the bridge. Forms are built out of wood and metal. Workers build forms at each end of the bridge. More forms are made in the middle.

Workers build the ends.

Machines pour concrete into the forms at the bridge's ends. The concrete becomes very hard. The forms are taken away. The hard concrete has become the bridge's **abutments.** Abutments are concrete shelves that hold up the bridge. One abutment goes at each end of the bridge.

11

Workers build the legs.

Workers pour concrete in the forms that stand in the middle. These forms make the bridge's legs. The legs are called **piers.** Piers hold up the middle of the bridge. A short bridge may have only one pier. A long bridge has many piers.

13

A machine lays beams.

A crane lifts long metal beams. It lays the beams across the top of the abutments and piers. Workers bolt the beams together.

Workers place bars.

Workers lay metal bars over the beams. The bars make a crisscross pattern. They make the bridge stronger.

17

Workers make the bridge's deck.

A truck with a long hose pumps concrete over the bars. Workers make the concrete flat and smooth. The concrete dries and gets hard. This smooth concrete becomes a road where cars drive. It is called the deck.

19

Workers add fences and lights.

Workers add concrete and metal fences on the sides of the bridge. Fences help keep cars on the bridge. Workers add lights to help drivers see at night. Then workers paint lines on the road. The lines tell drivers where to drive.

21

Cars drive across the bridge.

The strong bridge can easily hold cars and trucks and people. Bridges connect us to people and places.

Glossary

abutments (uh-BUHT-mehntz): ledges that hold up the ends of a bridge

beams (BEEMZ): long, thick pieces of metal

concrete (KON-kreet): a mixture of sand, stones, cement, and water

forms (FORMZ): structures used to shape concrete

piers (PEERZ): the legs or supports that hold up a bridge

24

Index

bars, 16, 18

cars, 6, 18, 20, 22

cement, 4

deck, 18

fences, 6, 20

legs, 12

lights, 20

machines, 10, 14

metal, 8, 14, 20